Mycelial Person

by Amanda Monti

Mycelial Person

Spore-radical contents:

+

Part I: the Mycelial Person

11

+

Part II: Florae filling w/ holes

37

+

Part III: the Weedy Tender

73

PART I

the Mycelial Person

Mycelium *(definition:)*

fungal underground networks, of which mushrooms are just a small visible part.

Person *(definition:)*

a human being regarded as an individual.

"No one fungal body lives self-contained, removed from indeterminate encounters. The fungal body emerges in historical merging - with trees, with other living and non-living things and with itself."

- Anna Tsing, *The Mushroom at the End of the World: on the Possibility of Life in Capitalist Ruins*

"What if we were to think of the person, like the fungal mycelium, not as a discreet, autonomous blobs, but as a bundle of lines, or relations along which life is lived?"

-Tim Ingold, "From Science to Art and Back Again"

"Nobody loves no one."

- Chris Isaak, "Wicked Game"

I wanted to become a mushroom.

I put an ad on Craigslist.

"Will you help me build a mushroom hat in exchange for a palm-reading?"

I made it explicit that there was no budget for this and all I had to offer was this divinatory reading, which I taught myself when I was fourteen because my mother was a witch with a pendulum and I had a heightened sensitivity for the effect of stories on hands and faces.

I didn't think much of it and would've felt lucky to hear back from even just one.

22 people replied.

some mushrooms

twinkle

in the dark to

lure insects

attracted to light

will crawl   such as firefly or flylarvae

who   between

round   round   round

over   under   the twinkling mushroom

&   &

in search

of   a

luminous lover

until they   realise that

oh...

it's just a mushroom

at this point   their little feet

are covered in

microscopic   sp*ores*

oh...

which the fly will   spread in service of the

dancing

fungus   *oh...*

fucking

At the time, I'd become lovers with my friend. It was very CASUAL, meaning that she was very casual about it and I was very much in love with her.

I tried to console myself by obsessing over the plant person from my local plant shop. I found out sun sign, politics, and hobbies through eavesdropping on other customers.

The Plant Person was a marxist Scorpio dancer, who always wore tank tops that revealed a tattoo of The Little Prince. Sometimes, The Little Prince cried droplets of sweat and I came to discern The Plant Person's scent as it mingled with the lucky bamboo and tiny palm trees. All I wanted that summer was to get a better glimpse at the inside of The Plant Person's palm and let the lines condensed into their skin reveal to us that these hands were, yes, wrinkly from the water, fragrant from the dirt, but also hands made to want me on a deep and intimate level.

Outside, the streets were creamy with micro milky ways as birds were still shitting on pavements and people were still talking about the weather like they had nothing to do with it.

The largest living creature on earth is mostly invisible to the human eye. *Armillaria Ostoyae* is a humongous fungus that covers almost four square miles below the National Forest in Oregon. Its network of roots, called mycelia, permeate the ground of the forest, weaving intricate lines into the deepest layers of the soil.

Wrinkles around skin, quietly telling their stories.

I honestly couldn't help reaching out after reading the ad description. I would love to help you make a mushroom hat.

Let me know,

B.

One night the doors of my biweekly sea creature movement appreciation improvisation club were locked. I had been forgotten.

I looked up at the sky and there was the mozzarella moon. I felt amorphous and expansive like pools of milk. I had nothing to do, nowhere to be. My night had begun with a failure, so I was once again free of having to have a productive time in New York.

Someone I knew had recently been awarded "The Best Expat in Madrid" award. The criteria seemed unclear, but some people just get an award for anything. It must be very exhausting. Winning is so very lonely.

I locked my bike and went to rite aid for pen and paper. Back by the scented candles, someone came up to me and asked, "Do I look like a person who should be chasing crackheads for the rest of my life?"
I didn't know what to say. I don't know what a person ought to look like, but I do think that mercy can be a climate.

When we were sixteen, we swallowed some ecstasy and sat by the riverbed, where we waited for the future to come get us. Instead, we were found by a man with a mini keyboard and a driver's license. He drove us to the suburbs, where we watched him eat fried noodles at 6am. Men will make you watch them eat fried noodles for the rest of your life if you don't step in.
Finally, he asked if we wanted to do crack cocaine with him in his bathtub. We politely refused and walked home for a long time.

There is a fungus that grows a spiky red stalk in autumn. People call it *Stinkhorn* because it oozes brown slime, which if received by the specialised cells high up in the human nose, will be recognized as 'shit' or 'rotten eggs.'

Some humans who have eaten Stinkhorn report that it leaves behind a sticky spore mucilage, which clings to throats and tongues tenaciously for hours and hours

            like an icky

                    kiss

Dear B.

Thank you so much for getting

in touch.

What material shall I get for our meeting?

A.

The pen I bought was purple, so I followed someone with a purple coat. They headed to a nearby bar. The person checking IDs called me "Sir," looked at me again with eyes lingering for microseconds on my chest, then apologized, as if it was their fault that gender was a fallacy.

The first time I noticed my breasts was in the murky water of my mother's bathtub. I was sad, but only because it was my understanding that I was now a "woman," and that this meant something less oceanic than what I had been. My mother looked like Princess Diana. Like Princess Diana, she did not seem as if she could be oceanic. She grew a thick curly chin hair, which I was in charge of plucking. "Life as a woman is tiring," she'd sigh each time it regrew.

Like her hair, my mother was resilient. When I was born, my mother smuggled me over the Italian border, because if you were a woman in Italy, you needed the father's permission to cross borders with a small baby and this signature she did not have. But my mother was determined.

"Andro," she said, "I will go."

And so she bought a ticket for the midnight train from Termini, cut a little hole in a tennis bag for me to breathe, and placed me inside. When they checked her at the borders, I remained quiet and invisible, like the secrets of a forest floor.

Where a mushroom ends or begins is hard to tell. Is it in the trees, the fruiting body, or the underground webs? And how easily it can seep into soil and pores and make them mushroom-like, too! It is something so many humans try to forget, the fact of our porosity.

You could make it out of old cardboard boxes, discarded hats, and fabric.

I have some old umbrellas I could give to you.

x

To strike up a conversation with the door person, I said, "protandric simultaneous hermaphroditism," which is a very long term to describe the simple gender switching lifestyle of the peppermint shrimp, who acquires certain gender traits, depending on its social situation.

The person checking IDs didn't show much interest and let me in. I sat down in a booth and decided to message one of the mushroom people.

I observed the crowds and wanted to see love. I tried to understand the rituals. My friend with the purple coat was now part of a semi-circle surrounding the DJ. There was liquid exchange in the form of tall alcoholic drinks and moist, fearless kissing. There was body language, slippery and soft. There was the performance of primary tasting organs being inserted into various holes. In my periphery I noticed a double-bearded man watching me watching. I began watching him watching me watching. I held his gaze to embarrass him, but feared it could be misread as flirtatious, so began thinking, *foul eggs, millionaires,* so as to communicate clear disgust through my eyes.

They started to play Chris Isaak's "Wicked Game." Hot, salty emotion flushed through my spine and out of my eyes. I had wrapped up two old lovers in this song. I did not think it was possible that a song could hold two people at once, but it did.

The secret was that you were the one who had to set boundaries.

Neither love nor Chris Isaak would do it for you. Both would just go on and on humming, like sirens.

There was also someone called Darwin and he had a daughter called Henrietta and Henrietta hated Stinkhorns. She set out for the forest dressed in her special hunting gear to collect the fungus and burn it, to protect the moral integrity of any nearby virgins, she said. The poor virgins. Did they gather in the forest? Did they smell the lingering Stinkhorn on her fingertips? Was it the scent that awakened them to her body?

Good luck with that, I wouldn't help you for a thousand dollars.

(anonymous)

The man's gaze and someone's shriek mating cry violently impressed onto me. I went outside and began taking notes. I had always hoped to write sentences so sad they could free me.

I wrote of the morning when we drove to the sea - the wind tore at the water and there was an impressive limestone arch that towered over the Atlantic Ocean. The power of the waves had eroded this rock and forged a hole through the middle, forming a natural threshold. Everyone seemed very happy on this side of the cliff. We should have been very happy, too. But I didn't see the sea that day, only a bunch of people looking at a portal, completely unmoving, while I longed to swim through it, to transform in it, entirely whole.

"...I just don't think it has to be," I said as I pulled the thick curly hair out of my mother's chin.

Mushrooms have a physical connection with rebirth. In order to keeping things moving, there are certain types of mushrooms who will decompose organic matter such as dead tree trunks into new nutrients. They accompany the tree into a new life, like the ferrymen of the Styx, or recycled melodies of pop music.

I think that I could be of valuable assistance to your project.

I don't need much more for payment than a fair assessment of my future. It doesn't have to be much of one, I am already fine with that.

Let me know if this could be arranged.

All the best,

Rich

"This world is only gonna break your heart," hummed a disembodied chorus as the camera panned across the music video for "Wicked Game," while my mum was on the phone to one of her long-distance lovers, while I was being indoctrinated into heterosexual desire by the music videos of the early millennium. White foam washed over the screen and revealed a sun-kissed Helena Christensen, bored and divine. As I sat through the entirety of the top 20, various expressions of masculine yearning condensed into a vague shape of a gaping wet mouth.

I wanted to be both Chris and Helena. I wanted the extremes of both boredom and pleasure. I wanted to be a man and woman and something else entirely. I didn't think there was a space in language for this, so I dug deep into the soil of where desire came from and covered it with earth. But desire is an organism that roots deep even if fruiting bodies are not immediately visible to the naked eye.

A *Volva* is a cup-like structure at the base of a mushroom that is a remnant of the universal veil.

The universal veil is a layer of tissue that completely surrounds the baby mushroom, making it look like an egg at first. After a while it drops, forming a halo around the mushroom.

A high number of mushrooms with Volva are deadly poisonous for humans so it is good to know your Volva in wild mushroom identification.

May I teach you knitting?

Sincerely, D.

"I was a very different person the last time I came to this city."

"Me too." This was a conversation I often had. When people asked after my heart, I began speaking of the water snake wiggler toy left over from the 90s. It was my erotic spirit animal. Neither in nor out, slippery and soft. An enigma. People did not congratulate me much, but that was fine by me.

I showed my companion the place you could get fresh cinnamon buns for fifty cents at 3am. I told them that my language was deeply confusing and that each word came with a gender. *Der Mond. Die Sonne.* I apologized for this arbitrary nuisance. Then we told each other tales of when we pretended to be boys and girls hitchhiking through mountains and I thought of how small bodies look on mountains, especially bodies of those we love, and about this blessing of a language that had the space to carry these small bodies, to carry them like ambiguous rocks, how terrifying that same language could be for its ability to crush whole cities under its weight.

I hadn't yet placed my companion as a future friend or potential lover, but because I gravitated towards the latter, I casually touched their elbow and commented on the fact that their creme sweater matched some of the baked goods. I was terrible at flirting and should have been born bioluminescent, lighting up for my loves, but for now this had to suffice.

Eventually, I walked them home. We looked at each other ambiguously, then some guy walked past and I missed the critical moment wherein a hug could have turned into a kiss. I walked to my station, where I missed the last train. I sat down in front of a baeckerei, unsure what my next move should be. I got out my pendulum: "Should I go back?" Anti-clockwise meant no. Clockwise meant yes.

Clockwise.

In some of Nordic mythology a *Vǫlva* is what you would call a witch. A Vǫlva was a female identified person who served as a link between the spirit world and ours, a threshold.

A high number of Vǫlvas were deadly poisonous for existing power-structures built on domination, so it was good to have one as a friend or mother.

....if it's not urgent I can help.

Sue

Der Stern, masculine.

However, in Italian: La stella, feminine.

The star.

Der Mond.

La luna.

The moon.

Die Sonne.   m

Il sole.

The sun.   M

                                                                          l
                                        S

                                                s   der

das

There are insects who love the smell of a stinky Stinkhorn. They dip their feelers deep down into the spore soup. Once they leave, they distribute the liquid, and that's how the Stinkhorn does sex.

like

this

or

this

or

here

this

this

or

Dear B,

Thats ok - let's meet after you get back ! How have you been otherwise ?

Take gentle care,

here

A.

Dear A.,
Just got back from Chicago and I've FINALLY got time to relax.
I'm totally down for resuming the hat project! Sometime this weekend ?

B.

Outside of the bar, I felt a profound connection to a nearby manhole. It shone oily with remnants of polluted summer rain. I always felt fondly about holes as they were an interface between worlds. Black holes.

Pendulums.

Pissing, sweating, kissing, lactating, praying, singing; we humans were essentially holes for stuff to pass through us. To prove my point, some asshole was yelling stuff opposite the street and this time I yelled back.

The door person outside of the bar turned to me, again.

In Italy, they had lots of shrimp. Had I been to Italy?
Yes, you could say that I was born-
He LOVED Italian women.

At this, I snorted. A little bit of phlegm hit the pavement, like slimy spores.
"Dude!" he said. I wondered how the gender switching shrimp might talk of itself in Italian. Unlike English, Italian has only one tense in which you can talk about yourself without any gendered signifiers, and that is the simple future:

"Andro"

I will go.

"What's important remains invisible to the eye," says the fox from *The Little Prince*. What he really means is: you cannot perceive spores with the human seeing organ.

Spores are little units of could-bes. They could become a complex organism. Or they could remain unicellular. Some spores can float up to mountaintops. Others can fly as high up as the stratosphere. In the stratosphere they get a little lonely because it's cold and not as good as a forest, so they huddle together. This is why when humans look up at the moon, sometimes we will feel all the things that
we could be.

Sue, did you know that spores at sea level can float up onto mountaintops?! How is Thursday for you?

A.

Rich, sorry to have missed you — Are you still in the city? How about Wednesday morning?

A.

Dear A,

I have been thinking about foam. Maybe if we cut some holes we can have your head be inside of it. Does that makes sense ? Or an old umbrella, of course.

B.

B.-

Funny enough I was born into a story with many holes.

A.

Split Gill Mushrooms have over 23,000 different sexual identities. The "sexes" don't really involve physical differences, either. They just are.

Der Mond. La luna. The moon.(moon). Planetary bodies, like shrimp, change gender in other languages. It's like little holes in the fabric. But holes can hold you, let you breathe.

B.!

This is the mushroom hat person from... almost a year ago!!!

This city just makes it impossible to meet... How have you been ?

I had a strange Winter. Falling in love, Falling out of love etc... in the end I made a mushroom hat. People helped me here and there - it felt important to just DO it, you know. Spring is coming, after all.

How did you know that there are only two types of arctic ice ?

It makes me think about the things that remain invisible to the eye.

Did you know that mushrooms travel as high up as the stratosphere?!

I'd still like to make another mushroom hat with you. I think it would be nice to meet after all this HIN und HER. Let me know what you think.

Unfortunately, I still don't have a budget but I can cook and do the palm reading.

Hope you've been well & all the best,

Amanda

PART II
Flora filling w/holes

I spent seven days navigating the city led by its weeds. I amplified each encountered weed with street chalk.I made an entry into my logbook, made poems from these notes, made friends from these encounters.

In this context I understand weeds as unwanted plants in a "human controlled" environment. I understand weeds as complicated kin.

.

Dumbo:

Close your eyes. You are surrounded by the ocean of the motorway.
Down by York Street there is a subway stop, where no one seems to care.

Upon entering you had to mark yourself as either top or bottom.
I admired the sense of certainty that my companion brought to
our brief relationship: *"I want this,"* she had said to me, waist high
in my kitchen. *"I want this,"* she had said to all the condiments of a
food truck. *"I want this,"* she now told the leathered butch with the
wristbands. I am terrible at decisions and asked if it was possible to
be both, trying to sound like I didn't care. Inside, people were
fucking one another in shadowy enclaves, some chained onto
latex-cushioned tables, others wriggling around like worms,
gleeful in the mud that was a basement of downtown Brooklyn. I
expected a force beyond my consciousness to lead me into a
constellation of bodies, naturally revealing the language of my
desire. Instead, I saw someone I knew from a conference in Rhode
Island. We made polite small-talk while a very small, masc-
presenting person got whipped by a femme teddy bear on a table.
It was nice

to watch   *as I am crouching by the weed a woman approaches and asks*
how in this   *me whether I am visiting and, these grow EVERYWHERE, but*
scenario   *she likes the colours, they remind her of spring, do I like it here?*

the masculinity of the person had nothing to do with
insertion, nothing to do with dominance. *"How is it going"*
"Have you found somewhere to live in the city" "I still have that
sculpture" Sighing. We turned to look at the climaxing Teddy.
"Hey, it was good to see you," said my friend. *"You, too." "Take*
care." I went to the toilette and googled signs that you are a
top, including but not limited to loving to be the big spoon. I
preferred forks so this wasn't helpful. *"Perhaps we are*
mystified by versatility," someone suggested in the commentary
section, *"because there is no hetero- trope built onto it,"* and I
took notes as light flickered on and off.

one another

making, breaking and changing origin

What are you drawing? It's a project about noticing, *oh, that's very interesting, what are you noticing?* This for example, *oh, that's very interesting,* would you want to join me? *No, thank you*

an

ice cream stick
nails

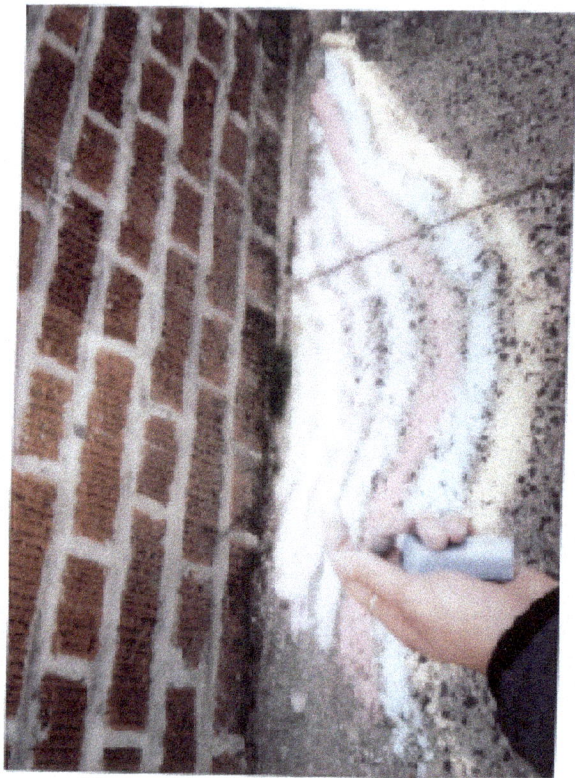

Ridgewood:

You are covered by the M train rattling above.

Spanish and American words leaking out of cars.

Next to trinity reformed centre church some fences light the sidewalk like a film noir.

This is where they have the Sauerkraut for 2$ and this is where I walked a woman

through the drizzling rain. This is where loud boys in flannel shirts

yell about Marx and where cute poets go for tea.

hair slicked back with vaseline lilac two piece. In my walk I am careful
not to name the plants I see. I know what it is like to be named. A flame
in the room with a famed lesbian author and her followers. We see each
other where it would be softest to the touch. A necklace made of
obscene fungal fingers

embraces their slender   *I colour the curb and hear someone,* but what is she
throat as they hold   doing? *They are gone before I can explain that I am*
                             writing a love letter.
eyes with me

exquisitely confusing air

thickens   like whipped cream   two ends of a rainbow   can see
themselves only through   the other

y, contra todo, nasce una   apola
it's probably about the seeing

Bed-Stuy:

Behind you, chubby cars. A discarded copy of Simone de Beauvoir
on the mighty stoops of a brownstone. *"Do we get to have take out?"*

Inside this crack, a strand of grass nestles in the opportunity presented
by the broken pavement. She had always wanted to be either an air
hostess or a princess Diana
double. Sometimes she tells of the *I become more confident in my drawings. I*
day wherein a man whom she had *take more time, linger for longer. Maybe this*
shared a Taxi with was struck by *is about pausing. I notice certain textures of*
the concrete. I begin to think a lot about my
her beauty and offered her *mother on my search for weeds. Her strange,*
unruly path. Her sighing. No one speaks to
employment at his airline. She *me today.*
was already alone then, but not as
a statement of power and self-actualization, more like a form of
self- protection. She refused. *"But maybe I could've managed."* I tell
her that I don't believe in men giving you things. *"Ach,*
Mandlchen..."

ache...

where are you from

can you hear me

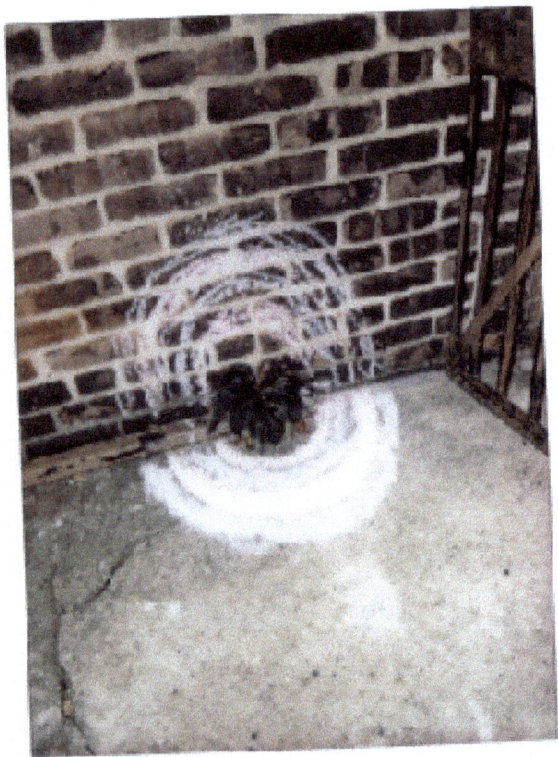

East Village:

"*Oh sorry.*" A girl smoking: "*Watch out.*" Purple
flags. American chatter. People with selfie sticks on
Broadway.

Slumbering beneath the mucous of this mural, a presence. Someone
from WWF stops me and tells me about the Koalas. I say thank you for
being out on the street. He says that "*our planet is dying.*" I know, I say,
I know. He asks me where my accent is from and, "*do I like it here?*" If
he means the planet, yes, there is beauty here.

Do you want to draw
around some weeds
with me? "*No, but
do you have some
money to spare for the
Koalas?*" I don't, I
am

*Today no one asks. Mostly people don't look. More
so, I can feel them not looking. This particular nation-
state teaches us how not to look. The children,
however, look. I begin to fill the shapes I draw with
my own shadow. Maybe this ritual is about absence.*

sorry. "*That's a shame,*" he says. Thank you, I say, everyone should be
out on the streets. I wonder how old he is, seventeen maybe. Where
was I when I was seventeen? Heartbroken, probably. Then his phone
rings, and he says, "*I gotta take this, nice to talk to you.*" That night I
dream of the Koalas and of Zadie Smith, who advises me to get involved
with local projects as we walk through a yellow village.

our planet is dying. it is not hours.

usually, the kids don't feel shy to talk to me. *Is this a telephone box for the spirits? Can I jump on this? Is this Hopscotch? Are you making a maze?* The parents pull them away before we can figure it out together.

except for once when a six year old human dressed entirely in yellow draws around the weeds with me until the whole curb is yellow too.

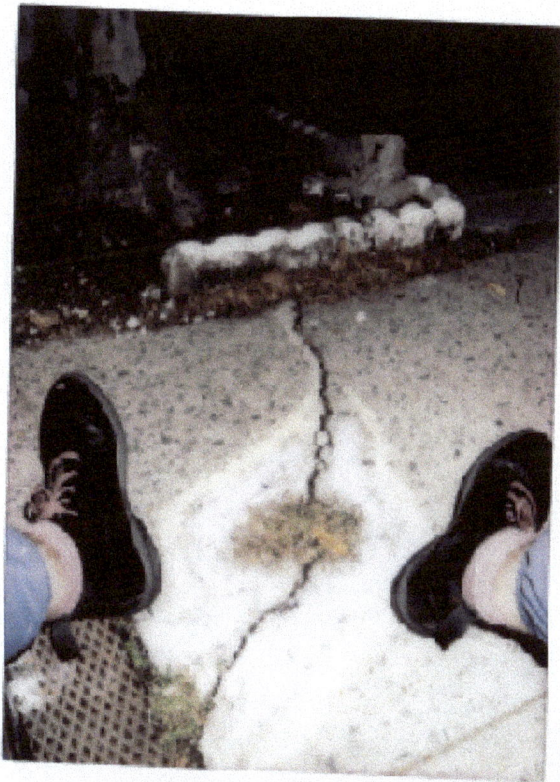

Greenpoint:

      Close your eyes. You are surrounded by distant hammering and Polish chatter. Underneath

     an old ice cream stick made tragic by winter you find her.

                                     i fell in love with a

                                                    a

a

cigarette butt grass

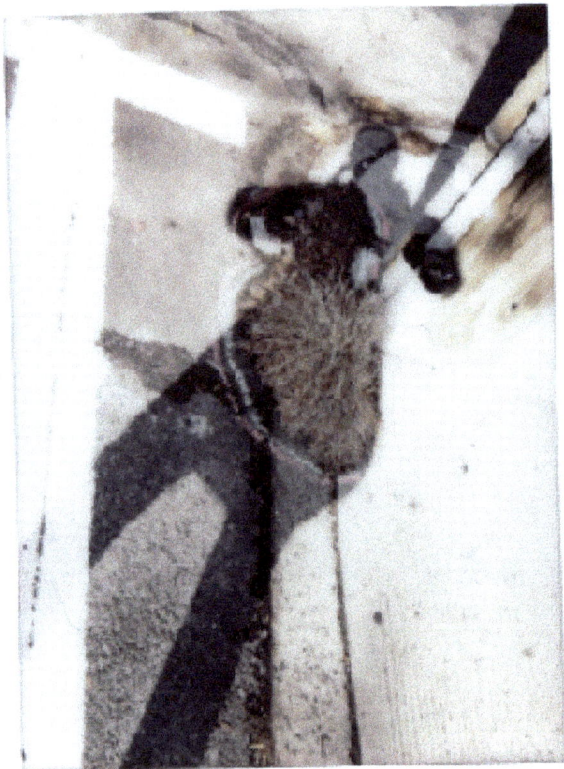

Vienna, 4th District:

Rascheln. Wölbungen. Outside the closed bookshop in the building you grew up in.
The familiar sound of Serbian families. One of the new E-Scooters is parked right
outside of the Papageno house, now rented out mainly to tourists.
Your mum loves it because it makes her feel like she's less here.

we used to live where mozart wrote the magic flute, I am long gone like
the bookshop and people come and go like it was the most normal thing
on this earth to travel to a place and just look at it with t-shirts and puffer
vests, the kids don't forget to smell the wooden chips, nutty and bitter
and I miss the duck shit in the ponds, polluted and thick like the lights in
the girls toilettes that shone purple so that you couldn't find your veins,
oh, they

don't *Someone asks* if this is street art? Is there an Instagram? *No, only*
gather in *memory.*

the Resselpark anymore, where they used to go in the warm
underground passage that connects church and opera house, where
Richard Strauss holds his violin, backside was embellished with needles,
where J and I went cack-cack-cack, even in the cold, but the whole
passage seems pointless now, because it's for people with puffer vests
and the people who are cold have been moved to colder places, but if you
try real hard you still get a whiff of cigarette stench coat and if you get
spat on at Mariahilfer cause you look a little queer, remember the tale of
Papageno, who was dressed like a bird and fails at every stage of the
magic flute but gets the happy ending for his good spirit, his lucky
charms, his perfect love. I certainly never thought of mozart in those
moments but maybe I should have, people eat chocolate with his face on
it after all and anyway I hear it doesn't happen all that often anymore, the
spitting I mean, the city has changed, e-scooters, no smoking, etcetera.

es sprießt

ein moosweiches

Verändern

caszauhause

moosweiches verändern das kind fragt gern

PART III

the Weedy Tender

glendale

you step
from the bus
into artemisia
she is moist
and carries
the scent of
sauerkraut
pointed leaves
she listens
slithers
into your
respiratory system
silvery undersides
armpit hair
and small purple flowers
how you remember

women

is how you remember
the concrete
lamenting

i miss the smell
of trees
of pulpy eyelash
in the spring

last season's   lametta

entangled in her

curb side

soot

c bergvall says space is doubt

remembering

is a sticky practice

like flushed up

six pack

yokes

space simmering

on a low, uncontrolled heat

its whole

mouth licking

future's teeth

artemisa vulagris

covered in cigarette ash

she stirs

her fingers

carefully

otherwise she will crack and

crystalize

in my reply

i water the weed

on my street fertiliser

from the corner shop

empty it into

the tongues of artemisia and she ripples

what they call invasive

species had travelled

as ballast

on ships

to counter-act the weight

of the bodies

violence

is

the blood of the world

not to notice

a weed

is another

concrete

crushes

at my most tender

touch

disposal plans

to open yourself
isn't enough

the blood oranges become
an expression of suppressed desire
baby, i want to li-

li-

literature you

to curb bush you

to go to some place quiet with you

to have you splayed open like a twenty dollar sandwich
and not wipe my mouth and ask for

extra sauce that

someone carelessly discarded on
the sidewalk
now it harbours
a dandelion
blow the
seeds
sponge like

structures will fuck

in the places where we listened

to each other

breathe

as old subway carriages

are thrown into the ocean

making artificial

reefs

pastoral

when they made the natural
world there were

no weeds

left, only lawns

and seminal texts on

the difference between

mucous and pus

all the while

pussy willow in the church

mouth

when i was

seven i saw the ecstasy

of saint teresa

lips parted in roses

and weeds? surely not

ineffective anti-

biotic travelled

through vessels

come, for

get me

for a moment

let's comb through the forest

re

member rolling

in the grass with me?

we carved initials

into a tree

trunk you were the first girl i knew

who owned a knife

while the grief we

collected grass and ticks with

our viscous call

it what you will

i was a weed

by which i mean

you took me

did you know that

knotweeds are weeds

that grow

and grow and

destroy real estate

and heal early stage

lyme

        disease ?

it was so obvious

all year round

we grow

back in the river

you took me

to the sewage, flashed your

blade to yelling neighbourhood

boys as we bathed in sulphur

skin slimy from

the first algae to inhabit these

waters, we kiss

mucous and pussy

in a poem

about the earth

temporal patterns other than the forward march

form taking the temporary manifestation of

red-soled ladies

form taking the temporary manifestation of

rubbing

form taking the temporary manifestation of

life-vests found underneath your seat

lined with precious feathers

a lover

fucks me hard

against a wall of

scratchy felt

she is pressing so hard

before i leave

"help yourself first before helping

others"

this is the top of the

pops

airport don't let

you enter into the humbling presence of limitation

thick clouds and a bunch of america, three grapes

powdered milk and a cake packed on a plastic tray

nothing good has ever come from

this

and the curb bush follows me

in the airplane a man takes off his trousers

touches his genitals through

thin leggings

all forms are

impermanent

except for this

my backside burns

i think of all the

pale boys on acoustic

guitars whining

unconsciously

squished into

one airplane toilette

it gives me

peace that it is

daytime soon

underneath an old ice cream stick

it's autumn

in the hotel of europe

there is a problem

in the back, front, and middle

yard there is

an infestation, they say

the knotweed,

is everywhere,

even in the

private property, they say

we cannot

say "we" anymore

only ew

ew, ew

ew

warm oats with switchy tendencies

cruising plant shops now

i get a coffee and piss into

the tulips i can't hold it in

this life came to me in a box

marked 'undisclosed

contents' i take an online

test to see what

kind of person and/or breakfast

i need

the person in the plant shop

displaying inconceivable amounts of patience

"sorry for pissing into the tulips"

"it is good fertiliser"

*

in the revolution

we will see each other for who we

really are but still have fun dressing up we

will be grown by tulips as big as trees

we will make another cum by barely even moving

we will attend to one another in a manner

that is generous and non threatening

and

*

the plant person leans in and whispers intomy

ear lobe "dig your

two fingers one

inch deeper

to test for moisture"

"so wet" i gasp

"eros is

just an issue of boundaries" says

the plant person

*

and the hawthorne tree blooms

drenched in light and our anger will

turn raw rain into forests

and we will unlearn to be lonely

*

i nod before

the plant person is even asking

at which the plant person begins

inserting

tulips

one by

one

"oh

my"

"all we

have

 is transcendence"

primordial broth recipe

if possible

crouch by a bush of mugwort leaking

out of the regulatory fiction

of a private park

all fences are intrusions and essentially

deformities

ask the mugwort for permission to take

a strand

tender now

hang it up to dry

for a couple of weeks

while she loves your

back

steep in hot water

space only exists

through brewing

stories

here, put your face

into my palm

just like this

motherhood

you bring it to your mouth

it is only fair that vitamins should be free

in this country

a swab to a swipe to a

subway stop

where the eyes of the people

are untouching themselves

again and again

while seeds have woven intricate webs

underneath the tracks

subversive disorders of

swiss cheese

how is it, ma'am?

i make it up

how hard i had loved you

on my face

if beauty is a gap

or a bridge depends

on who you pay

at sephora i test squeeze

exclusive purification

techniques

into my third eye point with

seventy two dollars worth of nano gold

drown you in hyaluronic

acid, retinol, planifolia

from madagascar

forever disguising

the fact of change

we demand exclusive vitamin c for all

i insert some lipgloss into my pelvic floor

squeeze, thief, squeeze

two for one at our own risk

an exclusive discount

at the spa on groupon

we get in a car where the man tells us

to be silent

he says be quiet

i don't argue, after all, i don't know

what it's like to be driving stupid people from and to the

spa again and again

the most money I have ever made

involved rubbing powdered turmeric

into the crack of my asshole

i would like to add that turmeric

grows in a rhizome

like artemisia vulgaris

difficult to tear out

because of its horizontal underground arms

reaching deeper than

you can stand before

looking away

turmeric lattes trending in 2019

i google "where does the US

get most of its turmeric from"

everyone looks happy "we love detox"

be quiet, he says again

and earth says, i'm bored of you people

and earth says, i am boiling

and he drops us off

with a few other fellow

wellness troopers

complimentary facial masks

pink two pieces

i am destined for

the hottest tent

where they put eggs into the ceiling

it is so hot that the eggs bake

and i sit and realise that my heart

exhausts me

i let go

form manifesting itself as

the experience of heat

form manifesting itself as

the experience of heart

ache

how hard

one can

love

words

"O"

"O"

"O"

alone

in the hot tent

maybe in the world

with the eggs

that bake

i reach

deeper understanding of

what it is like to be

bones

storing starches and proteins

and stories

earth says

attend

i can't exempt myself

from wanting

this shit pile

to contain love

when i get out of it

i am given an egg that was

baking with me

perhaps this is as close

as i will ever get

to the experience

of motherhood

Amanda Monti is a cross-disciplinary poet and translator born in Rome, raised in Vienna and based in Queens. They use playful research methodologies and a fantastically queer lens to explore ecology, language and desire through book objects, performances, sound walks, installations, divination, radio and pedagogy.

Amanda has been published through the Institute of Contemporary Arts London, The Poetry Project, McSweeney's, ExBerliner, Cuntemporary and Radiophrenia, amongst others. They have been the recipient of an Erasmus Scholarship and hold an MFA in Poetry & Literary Activism from the Pratt Institute in Brooklyn.

Mycelial Person is their debut collection and has been adapted for radio as *Spore Radical* (originally broadcasted on *Montez Press Radio*.)

Find them cruising at Ridgewood Reservoir or on the internet:

https://softie.space

Gratitudes

This work would not have emerged without the mentor-& friendship of Laura Elrick, gardener extraordinaire of many-a poetic weeds. Thank you for your joy, honesty, faith, care, and kinship throughout the making of this book.

I am grateful to Christian Hawkey, LA Warman, and Mirene Arsanios for their guidance, gentleness, and generosity; for keeping me both inspired and accountable throughout the process of figuring out this work.

Thank you to my dear friend Erika Hodges for their immaculate editing and world-making skills.

How would I have finished this book without Can Serrat Artist Residency? Thank you for gifting me with the three precious ingredients of time, mountains, and food.

And finally, thanks to Mum and to all the beautiful people I had the privilege of meeting and loving on the streets of Queens, the internet, nightclubs, sweaty basements, in the supermarket, by the sea, after the readings, dancing, at home, in forests.The soft idents you leave in my body allow me to re-fill them with poems.

www.ingramcontent.com/pod-product-compliance
Lightning Source LLC
Chambersburg PA
CBHW042336030426
42335CB00028B/3359